ASIA

BY MARY LINDEEN

CONTINENTS OF THE WORLD

Asia is the largest **continent** in the world. One-third of all the land on Earth is in Asia.

Asia is one of seven continents on Earth.

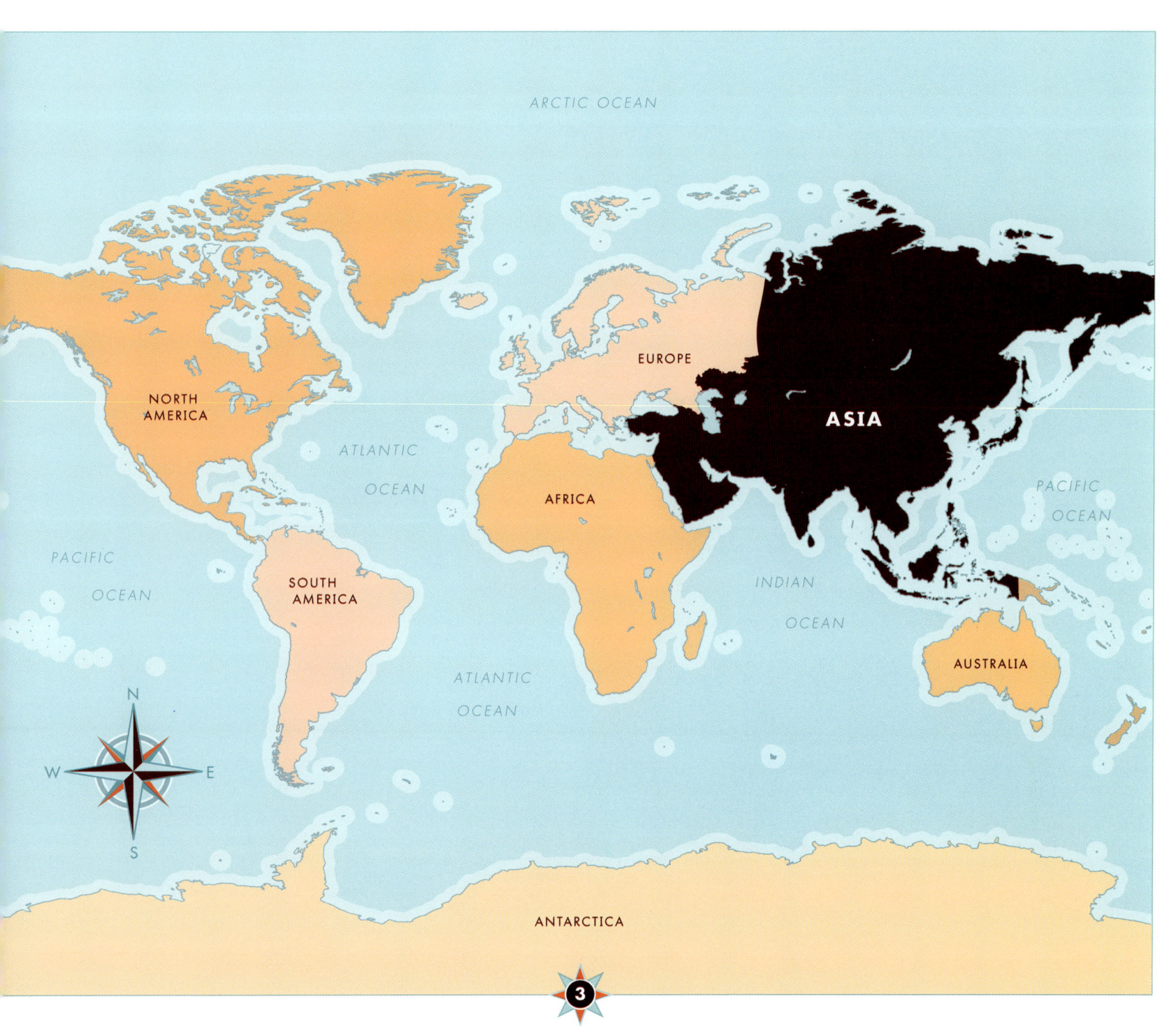

ASIA

Many **islands** in the Pacific Ocean are part of Asia. Volcanoes made these islands.

*Indonesia is an Asian **country**. It is made up of more than 17,500 islands.*

CONTINENTS OF THE WORLD

The highest mountain in the world is in Asia. It is called Mount Everest. Asia also has flat **grasslands**.

Mount Everest is 29,035 feet (8,850 meters) tall.

ASIA

There are deserts in some parts of Asia. In other places, there are farms and flowing rivers.

☜ *The Mekong River is in Asia.*

CONTINENTS OF THE WORLD

Most of the rice grown in the world comes from Asia. Tea and fruit are some of Asia's other crops.

Rice is planted in clumps. The field is kept wet for much of the time.

ASIA

Asia has many countries. The country of China has more people than any other country in the world. Other parts of Asia have hardly any people.

 Beijing is a busy city in China. Many people live and work there.

CONTINENTS OF THE WORLD

Asia is home to many kinds of people.

They enjoy different special events.

Many people **celebrate** the start of a new year.

These girls are dressed up for a New Year's *parade*.

ASIA

Many rare animals live in Asia. The giant panda and the orangutan are from Asia.

Wild giant pandas only live in China.

CONTINENTS OF THE WORLD

Asia has many buildings that are thousands of years old. It also has cities that are growing.

The Golden Temple in Japan is very old.

ASIA

Asia has so many different things to see and do. Where would you like to go first?

☞ *Many people visit the Taj Mahal in India, an Asian country.*

GLOSSARY

celebrate (SEL-uh-brayt): People celebrate when they do something fun for a special event. Some people in Asia celebrate the start of a new year.

continent (KON-tuh-nent): A continent is one of seven large land areas on Earth. Asia is a continent.

country (KUN-tree): A country is an area of land with its own government. China is a country in Asia.

grasslands (grass-lands): Grasslands are large open areas of grass where animals can graze. Some parts of Asia have grasslands.

islands (eye-lands): Islands are areas of land surrounded by water. The continent of Asia includes many islands.

parade (puh-RAYD): People marching for a holiday is called a parade. In some Asian countries, people celebrate the new year with a parade.

TO FIND OUT MORE

Books

Drevitch, Gary. *Asia*. New York, NY: Children's Press, 2009.

Hirsch, Rebecca. *Asia*. New York, NY: Children's Press, 2013.

National Geographic Society. *National Geographic Kids Beginner's World Atlas*. Washington, DC: National Geographic, 2011.

Web Sites

Visit our Web site for links about Asia:

childsworld.com/links

Note to Parents, Teachers, and Librarians: We routinely verify our Web links to make sure they are safe and active sites. So encourage your readers to check them out!

INDEX

animals, 17
cities, 13, 18
crops, 10
deserts, 9

grasslands, 6
islands, 5
Mount Everest, 6
Pacific Ocean, 5

people, 13, 14
rice, 10
rivers, 9
volcanoes, 5

Published by The Child's World®
1980 Lookout Drive • Mankato, MN 56003-1705
800-599-READ • www.childsworld.com

Credits: Ethan Daniels/Dreamstime.com: 4; FOTOAMP/Shutterstock.com: 8; Martinho Smart/Shutterstock.com: cover, 1; petoei/Shutterstock.com: 15; realmatee/Shutterstock.com: 19; Roop_Dey/Shutterstock.com: 20; San Hoyano/Shutterstock.com: 16; sunun/Shutterstock.com: 11; testing/Shutterstock.com: 12; Vixit/Shutterstock.com: 7

Copyright ©2019 by The Child's World®
All rights reserved. No part of the book may be reproduced or utilized in any form or by any means without written permission from the publisher.

ISBN HARDCOVER: 9781503824959
ISBN PAPERBACK: 9781622434169
LCCN: 2017960237

Printed in the United States of America • PA02372

ABOUT THE AUTHOR

Mary Lindeen is an elementary school teacher who turned her love of children and books into a career in publishing. She has written and edited many library books and literacy programs. She also enjoys traveling with her son, Benjamin, whenever and wherever she can.

On the cover: Rice fields can look beautiful in the morning light.